PCS
TO
PROSPERITY

BUILDING WEALTH
WITH EVERY MOVE

ASK ANTWAUN

PCS to Prosperity

Building Wealth With Every Move

ISBN: 979-8-9936505-1-7

Published by AskAntwaun Media

All inquiries: AskAntwaun@gmail.com
Printed in the United States of America

Book Design by Williams DocuPrep
www.williamsdocuprep.com

Table of Contents

Acknowledgements

To every Soldier, Sailor, Airman, Marine, and guardian who serves this book is for you. You've sacrificed time, comfort, and certainty in service to your country. Now it's time to serve your future.

To my family, thank you for your patience, support, and belief when I worked late nights building this vision. And to every client, colleague, and mentor who shared their stories, your experiences shaped every page.

Service doesn't stop when you hang up the uniform. Sometimes, it just changes missions.

Introduction

From PCS Orders to Profit Plans

For most of my career, every PCS meant one thing — *another restart*. New base, new address, new bills, same LES. I'd pack up my life, chase housing leads, and hope to find something close enough to base that didn't eat my BAH alive. I wasn't thinking like an investor; I was just trying to survive the move. I rented at every duty station. Why? Because that's what everyone around me did.

Owning a home felt out of reach, too complicated, and too risky, and honestly, I didn't want the responsibility while juggling the military grind. Like most service members, I was focused on the mission, not on money.

It wasn't until I separated from the Army that I realized something:

I had missed five opportunities to own property in

five different markets. Every PCS that I saw as a disruption could have been a step toward financial freedom. Each move could have built equity, created rental income, or opened the door to long-term wealth. But back then, I didn't know what I didn't know.

That realization lit a fire in me. I made it my mission to understand how military benefits really work and how they can be used to create wealth, not just cover expenses. That's what this book is about.

PCS to Prosperity is not about regret; it's about redemption. It's for every service member who thought they "missed their shot." The truth is, you can start today whether you're still in uniform or already out.

This book will show you how to:

- Turn each PCS into a stepping stone toward ownership.

- Research your next duty station like an investor, not a tenant.

- Use your BAH and VA Loan strategically to build equity.

- Create passive income that grows while you serve and after you separate.

You've been moving for the mission your whole career. Now it's time to move for your money.

Chapter 1

Understanding Military Pay

When most service members hear "PCS," they think stress, packing lists, and temporary lodging. They don't think wealth. But every PCS you receive is more than a change of duty station; it's an economic relocation opportunity.

Each move gives you access to a new housing market, a new BAH rate, and a fresh start. The problem is, most of us never see it that way because we're not taught to. For years, I didn't either. I went from base to base, renting because it felt simpler. I didn't realize that every rent payment was funding someone else's investment. Meanwhile, the same BAH I saw as "living money" could have been leveraged into a growing portfolio.

Here's the truth:

PCS isn't just a transfer; it's a transaction opportunity.

The PCS Wealth Gap

Every time you move, the government hands you three built-in wealth tools:

- **BAH (Basic Allowance for Housing)**—a guaranteed, tax-free housing allowance.

- **COLA (Cost of Living Allowance)**—extra pay to offset expensive locations.

- **Entitlement Reset**—a clean slate for new lending options and housing markets.

That's three financial levers, every single PCS. Here's where most service members miss out. They spend their BAH instead of building with it.

Suppose you're an E-6 with dependents stationed in Hawaii. Your monthly BAH is around $3,200. Over a standard three-year tour, that's roughly $115,000 in housing money tax-free.

If you rent, that money disappears. If you buy, that same amount could become equity, appreciation, or even passive income later. That's the PCS wealth gap, not in pay, but in perspective.

The PCS Rent-vs-Buy Reality

Let's look at a real-world example: a two-bedroom, two-bath townhouse near base, around 840 square feet, with two parking stalls and solid community amenities.

It's the kind of home many service members consider during a PCS move.

- Home Price: $540,000

- Down Payment: 0% (VA Loan)

- Interest Rate: 6%

- Monthly Mortgage: $4,191

- Tax Benefit: ≈ $816 per month

- Tax-Adjusted Payment: ≈ $3,374

- Total Payments (3 Years): $150,882

- Estimate Maintenance Costs: $8,545

- Estimate Appreciation: $58,708

- Principal Paid Down: $21,601

- Tax Savings (3 Years): $21,441

- Total 3-Year Cost of Ownership: $57,639

- Comparable Rent: $3,588 per month (2.5% annual increase)

- Total Rent Paid (3 Years): $129,176

Advantage of Buying vs Renting: ≈ $71,537

Breaking It Down

Over one PCS tour, here's what really happens with your money:

- If you rent, you'll pay about $129,000 in three years and have nothing to show for it. Every dollar goes to your landlord's mortgage, not yours.

- If you buy, you might pay more upfront, but your money starts working for you. After factoring in appreciation, principal pay-down, and tax savings, your true cost is about $57,600, which is less than half of what you'd spend renting. That's roughly a $71,500 swing in your favor in just three years.

Tax Benefits – Your Hidden Pay Raise

When you own a home, Uncle Sam rewards you. The money you spend on mortgage interest and property taxes can reduce your taxable income, meaning you keep more of your paycheck.

In this example, that's about $816 a month you get back in the form of tax savings. Think of it as a secret bonus, a "hidden BAH" most renters never realize exists.

Appreciation – Your Home Grows While You Serve

While you're focused on the mission, your home is focused on making you money. At around 3.5 % growth per year, that $540,000 home could be worth nearly $600,000 after three years, a gain of about $58,700, without you lifting a finger. That's the power of owning an asset instead of renting one.

Equity – The Payoff You Don't See Yet

Every mortgage payment chips away at what you owe and builds what's yours. After 36 months, you've paid down about $21,600 of your loan. That's real ownership, money you keep if you sell, refinance, or turn the property into a rental when you PCS again.

3-Year Rent Vs Buy Scenario

	Renting	Owning (VA Loan)
Monthly Payment	$3,588 (+2.5%/yr)	$4,191 ($3,374 tax-adjusted)
Total Paid (3 yrs)	$129,176	$150,882
Estimated Maintenance	—	$8,545

Estimated Appreciation	—	–$58,708
Principal Paid Down	—	–$21,601
Tax Savings	—	–$21,441
True 3-Year Cost of Ownership	$129,176	$57,639
Buyer Advantage vs Renter	—	≈ $71,537 gain

Golden Nugget Tips

- BAH is buying power, not spending money. Use it strategically.

- Your tax return is a refund on ownership. Put it to work.

- Every PCS move is an opportunity to keep your low VA rate and grow your portfolio.

Ask Antwaun Moment

"You're already paying a mortgage.
The only question is whose."

Pulling It All Together

Every PCS gives you a new BAH rate, a new housing market, and a chance to reposition yourself financially. You can either rent and restart, or you can buy, build, and stack equity with every move. That's the mindset shift that separates financial survival from financial growth. Because once you see PCS for what it truly is, a recurring opportunity to acquire appreciating assets, you'll never look at your orders the same way again.

Now that you see how each PCS can change your wealth trajectory, the next step is learning how to systematize it.

Buy → live → rent → repeat.

That's the PCS Wealth Formula, a process that even an E-5 can use to build a multi-state real-estate portfolio with just two or three moves.

Chapter 2

The PCS Wealth Formula

In Chapter 1, we reframed PCS as an opportunity. Now it's time to turn that opportunity into a system. PCS isn't just a cycle of packing and relocating; it's a rhythm you can use to create lasting wealth. The key is to treat each assignment not as a reset, but as a rotation through a new market that expands your portfolio.

The Formula: Buy → Live → Rent → Repeat

This is the foundation of the PCS Wealth Formula. With each move, you acquire a property using your VA entitlement, live in it during your tour, then convert it into a rental when you PCS. Then you repeat the process at the next duty station. Over a 10-to-12-year career, this pattern can build a multi-state portfolio that works for you long after you hang up the uniform.

Here's how to turn your PCS moves into long-term wealth, one station at a time.

Step 1: Buy Smart at Each Duty Station

The goal isn't to find your "forever home." It's to find a *strategic asset,* a property that meets your needs while you serve and performs as an investment afterward.

When selecting a property:

- Focus on location first: proximity to base, schools, and stable rental demand.

- Leverage your VA loan—zero down, no PMI, and lower interest rates.

- Buy at or below market value where possible. Equity on day one creates flexibility later.

- Avoid over-customizing. Think future tenant appeal, not dream-home perfection.

Each time you use your VA benefit, you're converting housing money (BAH) into ownership.

Step 2: Live with Intention

During your tour, your property functions as both a home and a wealth engine.

While you live there:

- Keep maintenance consistent. Small issues now prevent costly repairs later.

- Track your equity growth and home value annually.

- Reinvest savings from tax benefits into your next down payment or upgrades that boost future rent.

- If market conditions change, consider refinancing or using an IRRRL (Interest Rate Reduction Refinance Loan) to improve cash flow.

This is the *wealth incubation period;* you're living in your investment while it grows.

Step 3: Rent When You PCS

When new orders drop, most service members sell. That's where the cycle breaks. Instead, consider holding and renting. Turning your property into a rental does three things:

1. Creates cash flow. Your tenants now pay down your mortgage.

2. Builds equity and appreciation while you serve elsewhere.

3. Establishes a long-term income stream that continues after retirement.

If you bought strategically, your rent should cover your mortgage, taxes, insurance, and maintenance with room to grow as rents rise.

And if you used a low-rate VA assumption, your cash flow potential increases even faster.

Step 4: Repeat with the Next PCS

Each new set of orders gives you a clean slate to start again. Your entitlement resets once your prior VA-backed property is rented or refinanced conventionally. With discipline, even an E-5 or E-6 can accumulate two or three properties before retirement.

Imagine this:

- PCS #1 – You buy a townhouse near base.

- PCS #2 – You rent the first home and buy another at your next station.

- PCS #3 – You repeat again, creating three income-producing properties in three different states.

By the time you ETS or retire, your tenants have paid

down hundreds of thousands in principal, and your properties have likely appreciated by six figures. That's not luck; it's leverage and repetition.

Portfolio Power: How It Adds Up

Let's take a conservative scenario:

Three homes, each worth around $450,000, with 3% annual appreciation.

After 10 years:

- Portfolio Value: ≈ $1.48 million

- Equity (principal + appreciation): ≈ $400,000 – $500,000

- Monthly Cash Flow (post-service): $2,000 – $3,000

Those numbers aren't fantasy; they're built one PCS at a time.

Mindset Matters More Than Math

The biggest barrier isn't money; it's mindset. You don't need to be an investor to build wealth this way. You just need a plan, consistency, and the discipline to hold instead of sell. Every PCS is a reset button. Use

it wisely, and each move becomes a steppingstone toward financial independence.

Ask Antwaun Moment
"Every set of orders can be a
paycheck or a portfolio piece. You
decide which."

Understanding the PCS Wealth Formula is about
more than numbers; it's about awareness. Once you
recognize how your BAH, COLA, and entitlement resets create leverage, the next step is learning where
that leverage works best.

Not every market is equal. Some duty stations re-
ward appreciation, others cash flow. Some make
sense to buy, others to hold off and prepare. Your job
is to know the difference before you move because
when you understand the market you're stepping into,
you control the outcome before you ever unpack a
box.

Chapter 3

Market Scouting Before You Move

By now, you've started to see the PCS process differently. It's not just a relocation; it's an opportunity. So the next question becomes: How do you spot that opportunity before everyone else does? The answer is market scouting.

Think of it as reconnaissance. Not the kind you learned in the field, but financial reconnaissance. The same way every mission depends on solid intel, every successful move depends on understanding the terrain, the housing terrain. When orders drop, you shouldn't be reacting. You should already know which areas are growing, what homes are worth, and how far your BAH will stretch.

I didn't realize that while I was in uniform. I rented at every duty station because it felt easier. But after I separated from the Army and moved to Hawaii as a DOD civilian, I finally saw how much I'd missed. I ran numbers for the first time: BAH, average rents, and appreciation rates, and it was clear: I could have been building wealth every place I'd ever been stationed.

That's what this chapter is about: learning to scout the market before you move so that every PCS becomes a calculated step toward financial progress instead of another reset.

The Power of Early Intel

Military readiness means staying one step ahead. The same principle applies to your finances. If you wait until orders arrive to start learning a new market, you're already behind. Markets shift quickly. Interest rates rise, inventory tightens, and great homes disappear overnight. But early research turns pressure into preparation.

Let's say two service members are headed to San Antonio.

- One starts market scouting six months before orders, learns the average home price, and gets preapproved early.

- The other waits until arrival and finds out homes in their price range have jumped 5%.

The first service member walks into escrow with confidence. The second walks into another lease. Preparation doesn't just buy you options; it buys you time, and time builds equity.

The Five Layers of Market Scouting

To make market scouting effective, think of it in layers. Each one adds depth to your understanding of the housing landscape around your next base.

1. **Macro Trends**

   Start broad. Is the city or region growing? Are major employers expanding or leaving? Population growth, infrastructure projects, and job creation all drive appreciation.

2. **Military Economics**

   Compare your BAH and local rent averages. If the median rent equals or exceeds your BAH, ownership could be a smarter long-term move.

3. **Neighborhood Analysis**

   Focus on the five-to-fifteen-mile radius around your installation. Which neighborhoods have low turnover, good schools, and strong resale demand? Look at crime stats, average time on market, and school ratings.

4. **Property Type Fit**

   Decide what property fits your goal. Town-homes and condos are lower-maintenance but often carry HOA fees that reduce cash flow. Single-family homes appreciate faster and are easier to rent later.

5. **Exit Viability**

   Every buy should have an exit plan. Could you rent the home if you PCS again? Could you sell it quickly? Smart investors plan their exit before they enter.

Lessons From My Missed PCS Opportunities

During my years in service, I never viewed housing through a financial lens. I saw it as a necessity, not a strategy. Every home I rented whether in Georgia, Texas, or overseas went up in value. Looking back, all that BAH I paid could've been growing equity for me instead of someone else. That realization changed my mission. Now I teach others what I wish I'd known: that *each PCS is a business decision dressed as an assignment.*

The Market Match Matrix

Not all markets are created equal. Each base area falls into one of three categories, and knowing which you're entering determines your approach.

Types of Military Housing Markets

Market Type	Example Bases	Core Traits	Strategy
Equity Market	O'ahu (HI), San Diego (CA), Washington D.C.	High appreciation, limited cash flow	Buy for long-term value and resale profit
Cash Flow Market	San Antonio (TX), Fayetteville (NC), Oklahoma City (OK)	Lower entry price, strong rent ratios	Buy for rental income and hold long-term
Hybrid Market	Colorado Springs (CO), Tampa (FL), Las Vegas (NV)	Balanced appreciation and rent potential	Buy to live now, rent later

Tools and Tactical Resources

Here are the key resources I use and recommend to clients who are planning a move:

BAH & COLA Calculators

- Defense Travel Management Office (DTMO)

- MyPay (https://mypay.dfas.mil/)

Military.com BAH Calculator

- Market Research Sites

- Zillow & Redfin (for price and sales trends)

- Rentometer (for average rents and cash flow ratios)

- Niche.com (for community data and schools)

- PCS and local housing Facebook groups (for peer insights)

Economic Data Sources

- U.S. Census and Bureau of Labor Statistics (for population and job growth)

- Local Chambers of Commerce (for business development and infrastructure updates)

Gathering this intel six months before you move can save you thousands when you arrive.

Ask Antwaun Moment
"Financial readiness starts long before deployment orders."

Market scouting isn't complicated; it's discipline. The service members who learn to gather intel early stop competing for housing and start choosing their investments.

Every PCS gives you two paths: the familiar route of renting and resetting or the strategic route of researching and rebuilding. One drains your allowance. The other multiplies it.

Market scouting is how you identify the opportunity, but identifying it is only the beginning. Once you've learned how to find the right markets, the next step is knowing how to *move through them strategically*.

Every PCS gives you a new set of variables: different BAH rates, new duty stations, shifting home values, and unique lending limits.

What separates a one-time homeowner from a long-term investor isn't just knowing where to buy; it's knowing how to make the VA loan work for you every time you move.

Chapter 4

VA Loan Portability & Multi-Property Ownership

The VA Loan Isn't a One-Time Benefit

Most service members think of the VA loan as a one-and-done deal: you use it once, buy your "dream home," and that's it. That's the myth. The truth is, the VA loan is *portable*. It can move with you, be reused, and even be stacked under the right conditions.

That means every PCS doesn't have to end with selling your previous home; it can be the start of your next investment.

Mini-Glossary: VA Loan Terms Made Simple

Before we dive in, here are a few key terms you'll hear throughout this chapter explained in plain language.

Term	What It Means	Why It Matters
Entitlement	The dollar amount the VA guarantees to your lender on your behalf.	Determines how much you can borrow with zero down.
Partial Entitlement	When part of your benefit is tied up in another VA-backed property.	Lets you keep one VA home and still buy another.
Restoration of Entitlement	The process of resetting your benefit after selling or paying off a VA-backed loan.	Allows you to reuse your VA loan for your next home.
Intent to Occupy	VA loans must be used to purchase a primary residence you plan to live in.	You can later convert it to a rental after moving.
IRRRL	Interest Rate Reduction Refinance Loan.	A simple way to lower your rate or payment on an existing VA loan.

Antwaun's Story: The Missed PCS Opportunity

When I bought my first home in Hawaii as a DOD civilian, I finally understood the potential I'd overlooked for years.

During my time in the Army, I'd PCS'd multiple

times and rented every single tour. At the time, it felt like the easier option.

Looking back, I realized that if I had bought at even two of those duty stations and kept the homes, I could've owned multiple properties by now, all built on the same VA benefit I'd already earned. That's when I learned the truth: the VA loan isn't a single-use privilege. It's a lifelong leverage tool.

Lesson 1 – Understanding VA Entitlement

Your VA loan benefit is built on something called entitlement, which is simply how much the Department of Veterans Affairs guarantees to the lender on your behalf.

- **Full Entitlement:** If you've never used your VA loan or have completely repaid and sold your previous VA property.

- **Partial Entitlement:** When a portion of your benefit is still tied up in another home.

- Every time you buy with a VA loan, part of that entitlement is "in use." When you sell the home and the loan is paid off, you can request a **restoration of entitlement**, which fully resets your benefit, ready for the next purchase.

Ask Antwaun Tip:
Restoration isn't automatic. You have
to request it through the VA after the
sale closes. It can take several weeks,
so do that early.

Lesson 2 – Reusing Your VA Loan

Here's where most people miss the power of the program. You don't need to sell your current home to buy another with your VA loan; if you have remaining entitlement and can qualify for both mortgages, you can use it again.

Example: You buy a home in Texas for $300,000 using your VA loan. A few years later, you PCS to Hawaii, where the loan limit is higher. You can keep the Texas home as a rental and use your remaining entitlement to buy in Hawaii even with zero down, if your available entitlement covers it. That's how you start *stacking properties*.

Lesson 3 – Stacking VA Loans the Smart Way

Using the VA loan for more than one property requires discipline. Here's how to do it correctly:

1. **Know Your Remaining Entitlement.**

The VA will only guarantee a certain total amount (roughly 25% of the loan value). Ask your lender to calculate exactly how much is still available.

2. **Understand County Loan Limits.**

   Loan limits vary by location. A high-cost area like Oʻahu allows a larger VA-backed loan than, say, Fort Bragg. That difference can enable a second purchase.

3. **Qualify on Paper.**

   You'll need to meet standard debt-to-income ratios for both properties. Lenders may allow projected rental income from your first home to help offset that debt.

4. **Maintain Intent to Occupy.**

   The VA loan is for primary residences. You must intend to live in the new property, at least initially, to stay compliant.

5. **Plan for Exit and Restoration.**

   When you eventually sell or refinance into a conventional loan, request restoration of your full entitlement again; that's what keeps your VA benefit reusable for life.

Lesson 4 – Turning PCS Moves into a Portfolio

Each PCS gives you a choice: sell and reset or hold and expand. If you bought right, you can rent your old property out and move your VA loan to the next. Over time, every PCS can add another asset to your name.

Example:

- 2018 – Buy in Fort Hood for $250K (BAH covers mortgage).
- 2021 – PCS to Hawaii and keep Fort Hood home as rental ($350 cash flow).
- 2022 – Buy in Hawaii for $600K with partial entitlement.
- 2025 – Sell Fort Hood home and restore partial entitlement.

That's how ordinary PCS orders become an extraordinary portfolio plan.

Lesson 5 – Common Pitfalls and Misconceptions

- **"You can't have two VA loans."**

  False. You can if entitlement and qualifications allow.

- **"You must sell before using it again."**

  Not necessarily. Restoration isn't required if you still have remaining entitlement.

- **"The VA sets your loan limit."**

  Not anymore. Since 2020, there's no cap on VA loan size if you have full entitlement; limits only apply when part of your entitlement is in use.

Golden Nugget Tips

1. **Refinance strategically.** Use an IRRRL (Interest Rate Reduction Refinance Loan) to lower payments and free cash flow on older VA homes before reusing benefits.

2. **Keep meticulous records.** Every COE (Certificate of Eligibility) and closing statement will be needed for future entitlement restorations.

3. **Think portfolio, not property.** View each home as a stepping-stone in a long-term ownership plan, not an endpoint.

Ask Antwaun Moment

"Your VA loan isn't a single-use privilege; it's a lifelong asset. You earned it. Use it wisely, reuse it strategically, and let it build a foundation that grows with every PCS."

Understanding how to reuse your VA loan is what

opens the door to multi-property ownership, but knowing how to use that benefit is only half the equation. The other half is choosing *the right property* to begin with.

Every PCS market has good deals, bad deals, and smart deals. The smart deals are the ones that make sense for both living and investing. The challenge is defining your criteria.

In the next chapter, we'll break down how to create your PCS Buy Box, the framework that helps you evaluate school zones, commute times, rent ratios, and appreciation potential so every home you purchase fits both your lifestyle and your long-term wealth strategy. That's how you stop guessing and start buying with purpose.

Chapter 5

The PCS Buy Box

The Myth of "The Perfect House"

Every service member has felt it: that pressure to find *the* perfect house before Temporary Lodging Allowance (TLA) ends. The one with the dream kitchen, short commute, and enough backyard for the dog and the kids. But when you're building wealth through PCS moves, perfection isn't the goal; purpose is.

Your PCS home isn't just a place to live; it's an asset that needs to perform. That means learning how to define your "Buy Box," the set of specific criteria that makes a property work for your goals, your budget, and your next move. Once you know your "Buy Box," the buying process stops feeling like guesswork and starts feeling like strategy.

Lesson 1 – What Is a Buy Box?

A Buy Box is a decision-making framework. It filters every potential property through the same lens: Does this home fit the financial, lifestyle, and future investment goals I've set? Instead of chasing listings or relying on luck, you create a checklist that keeps emotions out and data in.

For PCS buyers, your "Buy Box" focuses on three things:

- 1. **Quality of life now:** school zones, commute time, daily convenience.
- 2. **Investment performance later**: rentability, appreciation, and maintenance costs.
- 3. **Ease of exit:** how simple it will be to rent or sell when it's time to move.

That's the PCS triangle: live well, earn well, exit easily.

Lesson 2 – The Core Criteria of a PCS Buy Box

Here's how to define your criteria and what to look for:

Category	Why It Matters	What to Look For
Location	Proximity to base and major employers drives demand.	10–15 miles from base, access to main highways, low commute times.

School Zones	One of the top factors renters and buyers value.	Above-average public school ratings (7+ on GreatSchools.org).
Rent Ratios	Determines if your home can cash flow when rented.	Aim for monthly rent = 0.8–1% of purchase price (e.g., $3,000 rent on $350K home).
HOA / Condo Fees	Eats into cash flow.	Keep fees below 15% of projected rent when possible.
Appreciation Potential	Long-term equity growth.	Look for areas with job expansion, new construction, or redevelopment projects.
Condition & Layout	Impacts maintenance and rentability.	Simple floor plans, newer systems, low deferred maintenance.

Once you define your "Buy Box," your agent can help you filter listings that meet those metrics.

Lesson 3 – Real Example: Building a PCS Buy Box

Let's imagine you're PCSing to Oʻahu as an E-6 with dependents. Your BAH is around $3,200 a month. You're open to buying but want a home that can later become a rental.

Your PCS "Buy Box" might look like this:

Criterion	Target	Reason
Commute	≤ 25 minutes to Schofield or Pearl Harbor	Keeps quality of life high
Price Range	≤ $650,000	Stays within 100% VA financing range
HOA Fees	≤ $150/month	Preserves cash flow potential
School Zone	Rated 7+	Protects resale and rental demand
Rent Ratio	≥ 0.8%	Ensures break-even or better as rental
Layout	3 bed / 2 bath / garage	Broadest rental appeal

With this box, you're not just shopping; you're strategizing. You can tell your realtor exactly what to target and what to skip.

Lesson 4 – Don't Fall for "Lifestyle Creep"

When you qualify for a higher loan amount, it's tempting to "stretch" for the nicer neighborhood or newer build. But remember: *what you can buy* isn't the same as *what you should buy*. Every dollar spent on square footage or finishes is a dollar that's not working for your long-term plan. Stay disciplined. Your PCS "Buy Box" is your compass, not your comfort zone.

Lesson 5 – The PCS Buy Box Formula

To keep things simple, use this formula:

"Buy Box" = (BAH + COLA) ÷ Market Reality. That means your budget should start with what you *already receive* tax-free (BAH and COLA) and then be adjusted by what the market tells you. If homes that fit your criteria are consistently renting for 1% of value and appreciating 3–5% annually, you're in a strong zone. If rent ratios are weak or prices are stagnant, it's time to adjust your Buy Box, not abandon your plan.

Lesson 6 – When the Market Doesn't Fit Your Box

Sometimes you'll scout a PCS market that doesn't align with your "Buy Box"—too expensive, too little rent return, or limited inventory. That's okay. You don't have to buy at every duty station.

That flexibility is what makes you an investor instead of a homebuyer.

Golden Nugget Tips

1. **Make data-based decisions.**

   Write your "Buy Box" down and review it with your agent before every showing.

2. **Think like a landlord before you buy.**

Ask, "Would this home rent easily to some-one like me?"

2. **Don't compromise on resale factors**.

Good school zones and locations near major employers protect your equity.

3. **Revisit your Buy Box, each PCS.**

Each market has different numbers. Your principles stay the same, but your parameters shift.

Ask Antwaun Moment
"Your Buy Box is your filter against emotional buying. It keeps you grounded, disciplined, and focused on what builds wealth not just what looks good on move-in day."

Every great investor starts with a clear box. Every great PCS homeowner becomes an investor by stick-ing to it. Defining your "Buy Box" helps you identify the *right kind* of property. But the next step is just as criti-cal: learning how to choose homes that *work for you now* and *work for you later*. Because owning the right property isn't enough if it can't transition easily from personal residence to rental or resale.

Chapter 6

Rent-Ready vs. Exit-Ready

Two Paths, One Purpose

When you buy a home during a PCS, you're really making two decisions at once:

- 1. How it fits your life today, and

- 2. How it performs when you move tomorrow.

Every service member who buys a home eventually faces the same question, "Should I rent it out or sell it when I PCS again?" The best answer comes from the day you *buy*.

If you purchase a home that's both rent-ready and exit-ready, you'll never be stuck. You'll always have options, and options are power.

Lesson 1 – What Does "Rent-Ready" Mean?

A rent-ready home is one that can easily transition from being your personal residence to an income-producing property.

Rent-ready properties have four key qualities:

- 1. **Desirable Location:** Near major employers, bases, or schools, somewhere tenants want to live.

- 2. **Functional Layout:** Nothing fancy, but practical: three bedrooms, two baths, parking, and laundry.

- 3. **Low Maintenance:** Durable finishes and updated systems that minimize repair calls.

- 4. **Strong Rent Ratios:** The projected rent should cover (or nearly cover) the mortgage, taxes, and HOA fees.

Rent-ready homes don't need to be glamorous; they need to be *profitable and predictable*.

Lesson 2 – What Does "Exit-Ready" Mean?

An exit-ready property is one that can be sold quickly and competitively when you need to.

That means:

- It's in a neighborhood with steady buyer demand.

- It meets lending guidelines for most loan programs (VA, FHA, conventional).

- It shows well and doesn't scare off appraisers or inspectors.

An exit-ready home is your safety net. If market conditions or life changes make selling the best choice, you can move fast without taking a financial hit.

Lesson 3 – Balancing the Two

Most homes lean toward one side or the other. Your goal is to find one that strikes the right balance.

Feature	Rent-Ready	Exit-Ready	Sweet Spot
Near Base	☑ Strong demand	⚠ Limited civilian resale	☑ Choose mixed neighborhoods
HOA Fees	⚠ Higher costs	☑ Maintains appearance	☑ Low but well-managed
Layout	☑ Simple and durable	☑ Modern and appealing	☑ 3-bed, 2-bath standard
Curb Appeal	⚠ Less priority	☑ Big resale factor	☑ Clean and low-maintenance
Upgrades	⚠ Focus on durability	☑ Focus on aesthetics	☑ Quality mid-range finishes

When you PCS, you want the flexibility to decide based on the market, not be forced into one choice because the property limits you.

Lesson 4 – How to Evaluate Rent-Readiness

Before you buy, run your property through a rent test:

- 1. **Check Local Listings.** Look at similar rentals on Zillow, AHRN, or Facebook Marketplace.

- 2. **Ask a Property Manager.** They can give you a realistic rent estimate, not the "wishful" number.

- 3. **Calculate Cash Flow.** Subtract your full payment (PITI + HOA) from projected rent.

- 4. **Stress-Test the Numbers.** Assume one month of vacancy per year and a small repair fund.

If you can still break even (or close), that's rent-ready.

Lesson 5 – How to Evaluate Exit-Readiness

For exit readiness, ask these questions:

- Is the neighborhood stable or growing?

- Are nearby homes selling quickly?

- Would a buyer need to do major updates?

- Does the home photograph well?

If you'd feel confident listing the property tomorrow, it's exit-ready.

Lesson 6 – The PCS Sweet Spot

The ideal PCS property hits the middle of both worlds: a home you're proud to live in that's also easy to rent or sell. That's what I call the PCS Sweet Spot: "Comfortable enough to live in, clean enough to sell, durable enough to rent."

When you buy at that intersection, you remove the fear from every future PCS. You're not trapped by the market you're positioned by design.

Ask Antwaun Moment
"Every PCS is a chapter in your wealth story."

You can either write it in ink with a property that keeps working for you or in pencil, where you're forced to erase and start over every move. Choose homes that keep writing your story, even after you've moved on.

Buying a rent-ready or exit-ready home is step one. Keeping it performing while you're stationed across the country or overseas is step two.

Chapter 7

PCS Property Management

From Landlord to Asset Manager

Owning a rent-ready home is one thing. Managing it while you're 3,000 miles away or halfway across the world is another. When I first became a landlord after separating from the military, I realized how quickly a good investment can turn stressful without systems in place. Missed rent, emergency repairs, or unreliable contractors can erase months of profit overnight. But the good news is this: You don't have to be a full-time landlord to manage your properties like a professional. You just need the right systems, people, and digital tools.

Lesson 1 – Treat Your PCS Property Like a Business

Your rental property isn't just "your old house." It's a business asset. Every dollar in or out affects your return on investment. That means structure and accountability are essential.

Here's how to manage like a pro:

1. **Create a Property Folder.**

   Keep all lease agreements, receipts, warranties, and inspection reports organized in a shared digital folder (Google Drive, Dropbox, etc.).

2. **Use Accounting Software.**

   Free or low-cost tools like Stessa, Avail, or RentRedi track rent, expenses, and tax deductions automatically.

3. **Schedule Maintenance.**

   Don't wait for something to break. Set a calendar for HVAC filters, pest control, and routine inspections.

4. **Communicate Professionally.**

   Treat your tenant like a client, not a friend. Keep all correspondence written and dated.

   Running your rental like a business means you'll always know its health and can hand it off easily to a manager or buyer when needed.

Lesson 2 – Deciding Between Self-Management and a Property Manager

The biggest decision every military homeowner faces is whether to manage their property personally or hire a professional.

Here's the breakdown:

Option	Pros	Cons
Self-Management	Full control, no management fees	Time-consuming, difficult from afar
Property Manager	Passive income, professional oversight	8–10% monthly fee, must vet carefully

If you're stationed locally and have time, self-management can save you money. But if you're moving out of state or out of the country, hiring a manager is almost always worth it.

Ask Antwaun Tip

"Interview multiple managers before choosing one. Ask about their average vacancy time, communication style, and how they handle repairs. A good manager can make or break your investment experience."

Lesson 3 – Digital Tools for Remote Control

Today, you can manage your property from any-where with a Wi-Fi signal.

Here are some tools that make it possible:

- **Rent Collection:** Avail, Apartments.com, Zelle for Business

- **Maintenance Requests:** Buildium, Hemlane, TurboTenant

- **Accounting & Reporting:** Stessa, QuickBooks Self-Employed

- **Video Inspections:** FaceTime, Google Meet, or virtual walkthrough services

- **Security & Monitoring:** Ring, Blink, or SimpliSafe for property oversight

These tools give you visibility and control without having to physically be there.

Lesson 4 – Building Your PCS Property Team

You can't manage from afar alone. Every great land-lord has a reliable *property team*:

1. **Property Manager** –The quarterback. Handles tenants, rent, and repairs.

2. **Handyman / Contractor** – Fixes issues quickly and fairly.

3. **Cleaner** – Essential for turnovers.

4. **Realtor** – Tracks market shifts and helps price future sales or new buys.

5. **Lender** – Advises when to refinance or pull equity.

Think of them as your remote control, each button handles a function so you can manage with confidence.

Lesson 5 – Handling the PCS "Surprise Call"

Every military family knows the drill: you're finally settled when the phone rings and you have new orders. If that happens and your property is already rented, don't panic.

Here's what to do:

1. Notify your property manager immediately.

2. Review your lease terms for notice and renewal windows.

3. Decide if you'll continue renting, sell, or transfer management.

4. Communicate clearly with your tenant; stability builds trust.

PCS moves will happen; chaos doesn't have to.

Lesson 6 – Protecting Yourself Legally and Financially

Distance doesn't exempt you from responsibility. Protect yourself before problems arise:

- Require renters insurance from all tenants.

- Maintain landlord insurance that covers both property damage and liability.

- Keep an emergency fund equal to 2–3 months of rent for unexpected costs.

- File taxes correctly and report rental income, but deduct eligible expenses (repairs, depreciation, mileage for inspections, etc.).

Being proactive here prevents headaches later.

Ask Antwaun Moment

"You can't control orders, but you *can* control outcomes."

Your PCS home should serve you whether you're living in it, renting it, or stationed on the other side of the world. With the right management systems your passive income becomes *controlled income*. Once your property is running smoothly and generating income, the next question becomes: "How do I use what

I've built to fuel my next move?" That's where equity comes in, one of the most powerful and misunderstood tools in military wealth building.

Chapter 8

Using Equity Between Moves

Equity: The Military's Hidden Asset

Every PCS move creates a new opportunity not just to relocate but to *reinvest*. And the key that unlocks those opportunities is **equity.** Equity is the difference between what your property is worth and what you still owe. Over time, it grows through two forces:

1. **Appreciation** – when the market value increases.

2. **Amortization** – when you pay down your mortgage balance.

Together, they quietly build wealth behind the scenes while you're focused on service, family, or your next move. But Equity only helps you *if you use it strategically.*

Lesson 1 – Don't Cash Out, Trade Up

When many service members see their property's value rise, the first thought is, "Let's sell and take the profit." That's the wrong move, at least if your goal is long-term wealth. Instead of cashing out, think of your equity as a reusable tool. You can leverage it to fund your next property while keeping the old one as a rental.

For example:

- You bought a home for $450,000.
- It's now worth $525,000, and your remaining balance is $400,000.
- That's $125,000 in equity.

You could sell and walk away with a chunk of cash, but you'd lose the asset, the appreciation potential, and the rental income. Or you could tap part of that equity to buy your next PCS property while letting tenants continue building wealth for you. That's trading up *not checking out.*

Lesson 2 – Understanding Your Options

There are several ways to tap equity between moves. The key is choosing the one that fits your timeline and goals.

Strategy	How It Works	Best For	Caution
HELOC (Home Equity Line of Credit)	A revolving line of credit secured by your home's equity.	Funding down payments or renovations.	Rates can fluctuate — use sparingly.
Cash-Out Refinance	Replace your current mortgage with a new one for a higher amount and take the difference in cash.	Consolidating high-interest debt or funding large purchases.	Increases your loan balance and resets your amortization schedule.
Bridge Loan	Short-term loan using your current home's equity to buy the next before selling.	Overlapping PCS moves or temporary transitions.	Usually higher rates and fees — plan to pay off quickly.

If your property is performing well as a rental, a HELOC is often the best choice. It's flexible, easy to manage, and lets you pull only what you need.

Lesson 3 – Timing Is Everything

The biggest mistake I see service members make is rushing to access equity right before or during a PCS. By that point, your focus is already divided between movers, orders, and timelines. For this reason, it's best to plan early. The best time to prepare for a future move is **one year before orders drop**.

Here's your rhythm:

1. **12 months out:** Get a CMA (Comparative Market Analysis) from a trusted agent.

2. **9 months out:** Talk to your lender about current equity, interest rates, and options.

3. **6 months out:** Decide whether to refinance, open a HELOC, or keep your loan as-is.

4. **3 months out:** Lock in your financing plan before PCS chaos begins.

That way, you are making calm, strategic decisions, not emotional, last-minute ones.

Lesson 4 – Don't Drain the Tank

Just because you *can* tap equity doesn't mean you *should*. Think of your home's equity like a retirement account. You can borrow from it, but every dollar you pull should have a purpose, not just potential.

Good equity uses:

- Funding your next PCS down payment.

- Making value-adding renovations.

- Consolidating high-interest debt to free up cash flow.

Bad equity uses:

- Vacations, cars, or short-term spending.
- "Investing" in things you haven't researched.
- Trying to time the market.

Every draw against your home should move your net worth forward, not sideways.

Lesson 5 – The Compound PCS Effect

Here's where it all comes together. When you strategically use equity between moves—buying, renting, leveraging, and repeating—you create what I call the Compound PCS Effect.

Each home you buy adds:

- Cash flow through rent.
- Equity growth through appreciation.
- Tax advantages through ownership.

Do that every few years, and suddenly your PCS orders become your investment schedule. This is how ordinary service members become extraordinary investors.

Ask Antwaun Moment
"Every move is a financial crossroad:
spend or leverage, sell or grow."

When you understand how to tap equity safely, your PCS becomes more than a relocation; it becomes a re-investment. Don't cash out your future for short-term comfort. Use what you've built to build the next step.

No one builds wealth in isolation. Your success in using equity or executing any PCS strategy depends heavily on who you surround yourself with.

Chapter 9

The PCS Investor's Network

Why Your Network Is Your Net Worth

Real estate is a team sport, especially when you're in uniform. Every PCS move puts you in a new market with new rules, lenders, and opportunities. You can either start from scratch each time or plug into a network that already knows the terrain.

That's the power of the PCS Investor's Network, a community of service members, veterans, agents, and lenders who understand how to turn mobility into momentum. When you move, your network should move with you.

Lesson 1 – Building Your Base Connections

Every duty station has a hidden network of people doing exactly what you're trying to do: buying, renting, investing, and building. You just have to know where to find them.

Start with:

- **Local Military Homebuyer Facebook Groups –** Many are base-specific and full of shared lessons, referrals, and property leads.

- **AUSA, VAREP, and Military Homeowner Events –** In-person connections often lead to trusted partnerships.

- **Your Local Realtor® Board –** Many agents specialize in military moves and understand VA loan nuances.

Ask Antwaun Tip

"Don't just collect contacts; connect with intent. Ask what others are doing, what's working in that market, and how you can provide value back."

Lesson 2 – The Power of a PCS Real Estate Team

You don't need to know everything; you need to know the right people. At minimum, your PCS Real Estate Team should include:

1. **VA-Savvy Realtor –** Knows base areas, VA loan rules, and rent-ready neighborhoods.

2. **Military-Friendly Lender** – Understands entitlement restoration, assumptions, and concurrent loans.

3. **Property Manager** – Keeps your asset producing when you move.

4. **Tax Advisor** – Helps track depreciation, deductions, and capital-gain timing.

5. **Insurance Agent** – Structures coverage for both occupancy and rental phases.

These aren't just contacts; they are your continuity plan between duty stations.

Lesson 3 – Digital Networking: Building Your PCS Rolodex

In today's market, relationships are portable. Use technology to create a "living" network:

- **LinkedIn** – Post about your moves, lessons learned, and what you're looking for next.

- **Instagram & YouTube** – Follow credible agents, lenders, and investors in upcoming PCS locations.

- **Google Drive or Notion** – Keep a digital log of contacts by city, base, and specialty so you can activate them fast.

Think of it as your **PCS Ops Order for Wealth—** every move, every connection, all in one place.

Lesson 4 – Networking by Contribution, Not Extraction

The biggest mistake investors make is showing up asking, *"What can you do for me?"* The right question is, *"How can I add value here?"* Maybe you share resources, introduce others, or document your journey publicly so new members learn faster.

Generosity is the ultimate credibility builder. The people who consistently give knowledge, leads, and time become the ones everyone wants to work with. Remember, reputation travels faster than orders.

Lesson 5 – Creating Your Own Micro-Network

If you can't find a network that fits, build your own. Start simple:

- Host a **monthly meet-up** for local service members interested in homeownership or investing.

- Create a **GroupMe or WhatsApp chat** for your base housing area.

- Start a **Facebook group** for your unit to share real estate insights.

Your first ten members will feel small until one of them closes their first property because of that chat. Then momentum takes over.

Ask Antwaun Moment
"Every property needs equity. Every investor needs connection."

When you combine both, you create a compound advantage: access to deals, advice, and allies who understand your mission. Your network turns every PCS from an isolated move into a coordinated campaign.

Concepts inspire, but stories transform. In the next chapter, we'll look at real service members who turned PCS orders into wealth ordinary people with ordinary paychecks who used these same principles to build extraordinary results. Because nothing proves the strategy works like seeing it in action.

Chapter 10

Real Stories of PCS to Prosperity

From Theory to Testimony

Up to this point, we've talked about strategy, numbers, and frameworks. Now let's look at how real service members turned those principles into action. These aren't celebrity investors or financial gurus they're Soldiers, Sailors, Airmen, and Marines who decided to play offense with their benefits instead of defense with their paychecks.

Each story follows the same formula you've learned:

- Understand the PCS opportunity
- Leverage the VA loan
- Buy strategically
- Manage smart
- Leverage equity forward

Case Study 1 – The E-6 Who Built a Portfolio by Accident

When Staff Sergeant Harris received orders from Fort Hood (TX) to Joint Base Lewis-McChord (WA), he and his wife bought a modest three-bedroom near base using his VA loan—$1,850/month mortgage, fully covered by BAH.

Two years later, new orders dropped, this time to Fort Liberty (NC). They debated selling but remembered the "rent-ready vs. exit-ready" rule. They rented the house for $2,100/month and used the small cash flow plus their tax refund to cover the next down payment.

Fast-forward five years:

- Property #1 in Texas now rents for $2,400 and is worth $80K more.

- Property #2 in North Carolina has $60K in equity.

- Combined, their net worth jumped $140K, not from raises or bonuses, but from keeping what they already owned.

Staff Sergeant Harris said, "I realized my PCS orders

were a buy signal. Every move, we kept one house and used the BAH from the next duty station to buy another."

Case Study 2 – The Dual-Military Couple Who Used Equity as Their Duty Station Bonus

Lieutenant Colonel Reyes and Captain Reyes were stationed at Ramstein AB in Germany when they decided to plan ahead. They bought a townhouse in Colorado Springs before PCS'ing, knowing it would appreciate quickly.

After three years, the home was worth $90K more than they paid. Instead of selling, they opened a HELOC for $70K and used $45K as a down payment on their next home in Virginia.

Today:

- Colorado property rents for $2,350/month, covering all expenses.

- The Virginia home sits $65K above the purchase price.

- They've since used the same equity strategy to buy a third home in Texas.

Their PCS cycle became a repeatable wealth machine: buy, rent, tap into equity, and repeat. The Reyes' said, "Our PCS used to mean starting over. Now it means leveling up."

Case Study 3 – The Veteran Who Started Late but Finished Strong

After 13 years of service, I left the Army and moved to Hawaii as a Department of Defense civilian. For over a decade of PCSes, I rented every time. It wasn't until after separating that I finally used my VA loan to buy. That purchase became my wake-up call.

That moment didn't just change my finances; it changed my mission. Now my goal is to teach others to do what I didn't: use their benefits while they still can. I missed years of ownership opportunities because I didn't understand the system. This book exists so you don't miss yours.

Lesson – The Common Thread

The first two stories have different ranks, bases, and budgets, but the pattern is the same.

- They used their PCS as a financial launchpad.

- They treated their BAH as a lever, not a lifestyle.

- They kept ownership through mobility.

That's how PCS becomes prosperity through strategy over circumstance.

Ask Antwaun Moment
"The difference between those who build wealth and those who don't isn't luck; it's literacy."

These service members proved you don't need perfect timing or huge savings to win; you just need a plan and the discipline to execute it every move. If they can do it with the same orders, pay grades, and time constraints you have, so can you.

Chapter 11

Golden Nugget Compilation

Your 20 Tactical PCS Wealth Moves

This is your quick-access field guide, a mission-ready checklist for turning every PCS into a step toward financial independence. These aren't theories or slogans. They're proven actions you can apply at your next duty station, no matter your pay grade or time in service.

1. **Every PCS is a Financial Reset.**

   Treat each move like a relaunch: new BAH, new COLA, new market. Reevaluate your housing and financial goals every time orders drop.

2. **Don't Spend BAH; Build with It**

   Your BAH isn't a housing allowance; it's investment fuel. Use it to build equity instead of paying rent.

3. House-Hunt Like an Investor, Not a Tenant

Evaluate properties based on rent potential, appreciation trends, and resale demand, not just comfort or commute time.

4. Own Before You Go

If possible, buy early in your PCS timeline. It gives time for equity to build and allows flexibility for rent or resale when orders change.

5. Use Your VA Loan Strategically

Know your entitlement limits, restoration process, and the ability to hold multiple VA loans when qualified. It's not one-and-done; it's reusable leverage.

6. Run a Rent vs. Buy Analysis Before Signing Any Lease

Compare projected rent costs against ownership benefits like principal paydown, appreciation, and tax savings. The math often surprises people.

7. Build an Emergency PCS Fund

Keep 3–6 months of mortgage and expenses

set aside. It protects your cash flow during vacancies, deployments, or delayed rent payments.

8. Rent-Ready Means Future-Proof

Choose homes with durable finishes, functional layouts, and easy maintenance so your property can become a rental without major upgrades.

9. Exit-Ready Means Option Control

Avoid homes that are too personalized or over-upgraded for the market. You want to sell fast if you need to not sit waiting for a perfect buyer.

10. Think Long-Term Market Compatibility

Buy in areas with both military and civilian demand. That keeps your property rentable or sellable even after you PCS away.

11. Manage Remotely with Systems, Not Stress

Use property management tools, auto-rent collection, and trusted vendors. You're not just a landlord; you're an asset manager.

12. Don't Drain Your Equity; Deploy It

Use HELOCs and cash-out refinances strategically to fund new properties or renovations. Only pull equity that moves your net worth forward.

13. Keep Great Records

Digitally organize everything: loan docs, leases, insurance, and receipts. You can't scale chaos.

14. Build Your PCS Real Estate Team

Identify your go-to agent, lender, property manager, and tax advisor at each new base before you move.

15. Network Before You Arrive

Join local Facebook groups, AUSA events, and real estate communities months before your PCS. Relationships open doors faster than pay stubs.

16. Treat Each Property Like a Command

Inspect, maintain, and track performance. Every home is a post in your financial chain of command.

17. Learn the Local Laws

Landlord and tenant rules vary by state. Know your rights and obligations; ignorance is expensive.

18. PCS with Purpose

Don't just follow your PCS orders; follow opportunity. Every move should either strengthen your cash flow or your equity position.

19. Teach as You Go

Share what you've learned. The more you teach, the more you solidify your own understanding, and you elevate the entire force.

20. Remember: Knowledge Is Leverage

You already have the benefits. You already have the paycheck. Now you have the playbook. The only thing left is execution.

Ask Antwaun Moment

"You can't control when or where the military moves you. But you *can* control how those moves shape your financial future."

If you follow even half of these strategies, you'll finish your service years with more than retirement points; you'll finish with assets that keep paying you long after you hang up the uniform.

You joined the military to serve a mission. Now it's time to serve your future.

Chapter 12

Conclusion: Mobility is Leverage

You Move for the Mission. Now Move for Your Money.

For most of your career, PCS orders have felt like interruptions. They've dictated when you pack, where you live, and how often your kids switch schools. They've disrupted routines, friendships, and stability. But what if those same orders were the key to your financial freedom?

What if, instead of viewing relocation as a reset, you saw it as a *reinvestment*? That's the mindset shift that changed everything for me, and it can change everything for you, too.

Lesson 1 – The Wealth Was There All Along

The military has already given you the tools:

- Guaranteed pay.

- Tax-free allowances.

- Access to the most powerful loan program in the country.

- A built-in schedule of relocations that naturally forces you to buy, hold, and diversify in multiple markets.

You don't have to invent a strategy; you just have to recognize one that's already in motion. Each PCS is an automatic opportunity to *build*, not start over.

Lesson 2 – Leverage Isn't Just Money. It's Mindset.

Leverage isn't about borrowing; it's about *positioning*. It's using what you already have to create something greater. BAH isn't just housing pay; it's seed capital. Equity isn't just home value; it's your future portfolio. Mobility isn't a burden; it's built-in diversification.

When you adopt this mindset, your career, no matter the branch or billet, becomes a wealth engine powered by movement. You've been leveraging your time, skills, and service for the mission. Now it's time to leverage them for your family's legacy.

Lesson 3 – Wealth Is Built in Motion

Military life is full of transition. The question isn't *if* you'll move; it's *how you'll move*. If you buy with intention, manage with structure, and plan with purpose, every set of orders moves you forward, not sideways.

Wealth doesn't happen by staying still; it happens in motion. The same mobility that once felt like instability becomes your greatest asset.

Ask Antwaun Moment

"If you've made it this far, you already think differently than most."

You're not just a service member following orders; you're a strategist building wealth through movement. The next PCS isn't a challenge; it's your next investment briefing.

The next set of orders isn't uncertainty; it's opportunity. So when that email hits your inbox, "New assignment," take a deep breath and smile. Because now you know the truth: Mobility is leverage. Movement is wealth. And your mission is prosperity.

Final Words

Thank you for letting me walk this journey with you.

I wrote this book not just as a real estate agent, but as a veteran who learned the hard way what I wish someone had told me years ago. If this book gave you clarity, share it with someone who needs it—a young service member, a family PCS'ing for the first time, or a veteran starting over. That's how we build not just wealth but wisdom across the force. And remember you're never in this alone.

Need real estate answers? **Just Ask Antwaun.**

Epilogue

The Mission Continues

When I first joined the Army in 2000, I didn't know what financial leverage was. I just knew how to work, serve, and move when told to move. Thirteen years, five PCSes, and two deployments later, I looked back and realized something powerful: I'd done everything right as a Soldier, but everything wrong as an investor.

I rented every time. I saved but never built. I worked hard, but my money didn't. This book is what I wish someone had handed me at Fort Hood, Schofield, or Benning: a playbook for using every move to move forward. Because it's not just about owning property; it's about owning your financial future.

Today, my mission is simple: To teach service members, veterans, and military families how to turn their benefits into wealth one PCS at a time. If this book helped shift your mindset, don't let it end here.

There's a community waiting to help you take the next step. Connect, learn, and keep asking questions. You already know the name. You already know the mission.

Need real estate answers? Just Ask Antwaun.

About The Author

Antwaun Hill is a U.S. Army veteran, real estate professional, and educator based in Hawaii. After 13 years of active-duty service, he transitioned to civilian life and discovered how misunderstood military financial benefits truly were, including his own.

That realization sparked a mission: to help service members, veterans, and military families use the VA loan, BAH, and PCS opportunities to build lasting wealth. As the founder of **Ask Antwaun**, he's become a trusted voice for military real estate education, blending strategy, storytelling, and empowerment.

Through his brand and book series, *BAH Means Buy A House*, Antwaun continues to teach one core message: "You already have the benefits. You just need to learn how to use them."

Follow him on Instagram **@AskAntwaun**

Contact: **AskAntwaun@gmail.com**

Need answers?
Just Ask Antwaun

Ready to put your plan in motion? Scan below to set up your free strategy session with Antwaun Hill.

Other books in this series

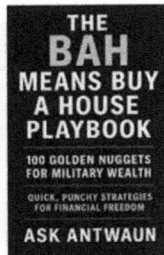

www.ingramcontent.com/pod-product-compliance
Lightning Source LLC
Chambersburg PA
CBHW071504210326
41597CB00018B/2679